SHARK BAY

DISCOVER MONKEY MIA & OTHER NATURAL WONDERS

DEPARTMENT OF CONSERVATION AND LAND MANAGEMENT

Published by:
Dr Syd Shea, Executive Director,
Department of Conservation and Land Management, 50 Hayman Road, Como, W.A. 6152.
Text:
Barry Wilson
Photographers:
Auscape International
Bill Bachman
Battye Library
Clay Bryce
Gerhard Freudenthaller
Robert Garvey
Lochman Transparencies
Greg Keighery
Mike Morcombe
Photo Index
Barry Wilson
Bert Wells
CALM Publications:
Sweton Stewart
Marg Bryant
Craig Garratt
Acknowledgements:
CALM wishes to acknowledge everyone that helped with this publication, including special assistance from Paddy Berry, Ken McNamara and Alex Baynes of the W.A. Museum and John Sellenger, President of Shark Bay Shire.

ISBN 0 7309 2532 3

CONTENTS

Discover Shark Bay

Shark Bay is a corner of the world which nature and history have endowed with a special magic.

Although the land appears bleak and arid, the eye cannot escape the breathtaking colours and distant brilliant vistas.

Red windblown sand dunes, spectacular limestone cliffs, gypsum rich claypans (birridas), pristine white beaches and sparkling green water are carved into an ancient and varied landscape.

Numerous bays, inlets and islands scattered throughout shallow waters interspersed with sand banks and seagrass meadows support a profusion of aquatic life.

At Shark Bay you can experience unusual natural beauty, interact with dolphins that daily come to the shore, explore an aquatic paradise, fish in warm turquoise recesses or wild open ocean or walk along beaches covered 7 to 9 metres deep in shells.

Welcome to Shark Bay. We invite you to a region of unspoilt beauty, a landscape of awesome grandeur. A rare experience awaits you.

Above: Big Lagoon at dusk.

Upper Right: Zuytdorp sea cliffs named after the Dutch ship "Zuytdorp" wrecked in 1712.

Upper Middle: "The Old Pearler" restaurant in Denham made of solidified shells.

Upper Far Right: Entrance to Big Lagoon

Lower Right: Meeting a friend

Lower Middle: Salt pans (birridas) stretch across wide coastal flats.

Lower Far Right: Experience beauty and freedom at Shark Bay.

An Explorer's Paradise

It was on these arid shores nearly four centuries ago that the first documented landing by a European occurred on the Australian continent. Little did Dirk Hartog realize when he stepped ashore at Shark Bay and described it as the "useless southland" how extraordinary an environment it really was.

The region is the sole habitat for a host of rare plants and animals and some of the islands provide the last refuge for several mammals and birds facing extinction.

The marine environment of the bay is unlike any other. Because of an unusual combination of geological, biological and climatic factors, the most isolated basins are highly saline and these areas support a restricted but peculiar flora and fauna, including the world-famous stromatolites.

Elsewhere in Shark Bay, the shallows are dominated by vast seagrass meadows providing nourishment and nursery to dugongs, turtles, prawns, scallops and fish of the bay.

Indirectly, the seagrass in Shark Bay has led to the development of banks and flats of sedimentary deposits which have no equal. As well as their geological interest, the wide intertidal flats which characterize the shores of Shark Bay create the habitat of a unique community of burrowing molluscs and other invertebrate species.

These outstanding features make Shark Bay special.

Burrowing Bettong, Shark Bay Mouse, Banded Hare-wallaby

Green Turtle

Sandhill Frog

Dragon Lizard and Honeyeater

Note — The Stromatolites picture (above) has been inadvertently transposed with the Sandhill Frog. Stromatolites are found at Hamelin Pool.

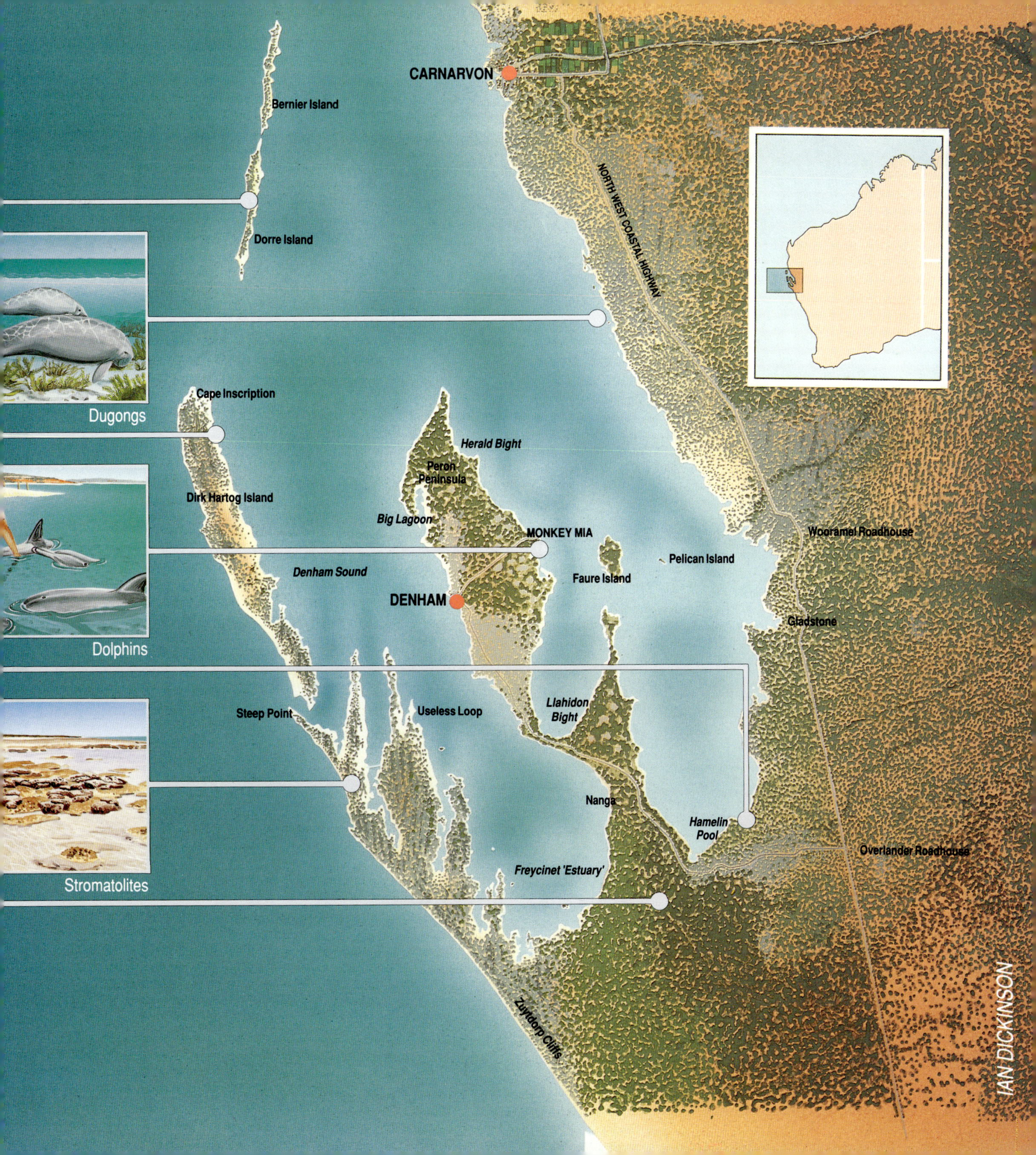
CARNARVON
Bernier Island
Dorre Island
NORTH WEST COASTAL HIGHWAY
Dugongs
Cape Inscription
Herald Bight
Peron Peninsula
Dirk Hartog Island
Big Lagoon
MONKEY MIA
Pelican Island
Wooramel Roadhouse
Faure Island
Denham Sound
DENHAM
Gladstone
Dolphins
Llahidon Bight
Steep Point
Useless Loop
Nanga
Hamelin Pool
Overlander Roadhouse
Stromatolites
Freycinet 'Estuary'
Zuytdorp Cliffs
IAN DICKINSON

History and Heritage

People have lived in Shark Bay for many thousands of years. Long before European contact, aborigines lived in the region and an archaelogical record of cave shelters and shell middens can still be seen today.

The first known landing of a European on Australian soil was made by Dirk Hartog in 1616 at Cape Inscription, at the northern end of Dirk Hartog Island. Willem de Vlamingh revisited Hartog's landing site in 1697 when searching for survivors from a missing ship.

William Dampier, the first English visitor to Australia, landed in the same area in 1699. He spent seven days in Shark Bay examining the shore and his name of Shark Bay remains today. A Frenchman, Francois St. Allouarn, also landed and made a formal claim to this territory for France in 1772.

Following these European explorers there were several French and English scientific expeditions to Shark Bay, perhaps the most famous being the French Baudin Expedition of 1801. During this expedition the French naturalist Francois Peron collected many biological specimens, which are preserved in the Natural History Museum in Paris. Most of the islands and bays are named after members of this expedition.

The first pastoralists arrived in the 1860's, about the same time as the pearl traders.

By the early 1870's, small settlements were scattered along the shoreline and pearling attracted many Malays, islanders and Chinese to the region. After 70 years of trading, the Depression in 1930 caused the pearling industry to close.

Subsequently, fishing became the mainstay of the district and a cannery and processing works were established in 1912.

Today, Shark Bay thrives on tourism, fishing and wool. A solar salt project at Useless Loop has produced another viable commodity contributing to a growing export trade.

Above: William Dampier, the first Englishman to land in Australia. *(Photo Courtesy of Battye Library, 5120P)*

Right: A fisherman's basket: the bay abounds in snapper, cod, groper, prawns, scallops, whiting and mullet. Off - shore are marlin, sailfish and mackerel.

Opposite Page: A bird's eye view of a stunning mosaic formed by the solar salt works at Useless Loop .

Jewels of the Desert

Shark Bay offers an exceptional range of plants with over 150 species of scientific interest occuring in the area. In spring, like most of coastal Western Australia, the area bursts into a kaleidoscope of colour.

Here is the boundary between two major botanical regions - the wattle dominated vegetation found in the Pilbara (Eremaean Province) and inland Western Australia and the eucalypt dominated bush of the South West. (South West Province).

Below: The aromatic Shark Bay Rose *(Diplolaena grandiflora)*

Below Right: A coastal daisy *(Senecio lautus)* growing on windswept cliffs has adapted to withstand high temperatures and long dry spells.

The vegetation on the eastern shore of the bay is mainly shrubland dominated by bluebush or acacia with patches of sandplain. Near the shore there are samphire flats and salt flats with algal mats, while the shore itself has a narrow zone of mangroves. The Peron Peninsula, and much of Nanga Peninsula, are also dominated by Acacia, with some mallee and sandalwood trees.

On the southern shore of Freycinet Inlet and southwards to the Murchison the vegetation is very different. It belongs to the species-rich South West district and is the northern limit of many South West species and genera. Much of this area is dominated by unique 'tree-heath' with eucalypt and banksias. The flora here includes rare and endangered

species and some are not yet described and named.

The vegetation of the Edel Land peninsula and Dirk Hartog Island is a curious mixture of both South West and desert region plants.

Far Left: This spectacular banksia *(Banksia sceptrum)* flowers in summer and provides nectar for honeyeaters and insects. At 30cm, its cone flower is unusually long.

Left: Stunted shrub vegetation including species of acacia.

Below:A wasp feeds on native myrtle *(Thryptomene baeckeacea)*.

Region Rich in Wildlife

For such a dry land the Shark Bay district is surprisingly rich in animals. The area is home to at least 10 species of mammals (some rare and endangered) and 98 species of reptiles.

This species-richness is due to the variety of land habitats present and the fact that this is an overlap zone between the fauna of the winter rainfall area of the South West and that of the arid interior zone.

Among the 100 or more species of bird found at Shark Bay one of the most interesting groups are the fairy wrens. The Black and White Wren (a sub-species of the widespread White-winged Fairy Wren) is only found on Dirk Hartog Island and Barrow Island. The Thick-billed Grass Wren once common in south Western Australia is now found in only two small populations, one of which is on Peron Peninsula.

Terns, gulls and pelicans nest on the smaller islands. At 'Shagmire', south of Louisa Bay on Dirk Hartog Island, there are large nesting colonies of pied cormorants. Substantial numbers of Osprey, White-breasted Sea Eagles and Wedgetail Eagles can also be seen throughout the Bay.

One of the strangest creatures in Shark Bay is the rotund Sandhill Frog which lives most of its days buried in the sand on the dunes of Edel Land. It emerges to feed on insects when it rains or in the night dew. It does not have tadpoles. The young frogs hatch directly from large eggs laid buried in the sand. Although this frog seems to be quite common in Edel Land, it has a restricted distribution and therefore needs special protection.

Opposite Page Top: Tree Dtella gecko *(Gehyra variegata)* is found in low shrubland in Shark Bay.

Opposite Page Below Left: A harmless Thorny Devil *(Moloch horridus)* which eats only ants.

Opposite Page Below Right: Majesty at rest; Osprey *(Pandion haliaetus)*, a fish eating bird of prey.

Left: The rare Black and White Wren *(Malurus leucopterus)*.

Below: Sandhill Frog *(Arenophryne rotunda)*.

On the Brink of Extinction

Since the time of European settlement and particularly during the past half-century, many middle-sized native mammals have become extinct on the Australian mainland. The Shark Bay island reserves of Bernier and Dorre Islands are now the last refuge of a number of species once widespread on the continent.

A total of seven native ground mammals has been found on these two arid islands. Four of these are totally extinct on the mainland - the Western Barred Bandicoot *(Perameles bougainville)*, the Banded Hare-wallaby *(Lagostrophus fasciatus)*, the Burrowing Bettong *(Bettongia lesueur)* (which also survives on Barrow and Boodie Islands), and the Shark Bay Mouse *(Pseudomys praeconis)*. Two other species of mice are still widespread on the

Above: The rare Shark Bay Mouse *(Pseudomys praeconis)* restricted to Bernier Island survives on leaves, stems and flowers.

Top Right: Hostile shores of Dorre Island.

Right: A Burrowing Bettong *(Bettongia lesueur)* was previously found in all states except Queensland and is now found only on Bernier, Dorre, Barrow and Boodie Islands.

mainland. The Rufous Hare-wallaby *(Lagorchestes hirsutus)*, just survives in the northern desert.

Bernier and Dorre Island are long and narrow, with a total area of only 10 500ha. They are covered with low shrublands, and are subject to salt-laden spray and frequent strong winds. They seem an inhospitable last refuge for mainland mammals.

The survival of these native mammals is precarious at best. Feral goats once present on Bernier Island have now been eradicated and, so far, foxes and cats have not reached either island. Fire, severe change in climate and introduced predators could threaten their continued survival.

However, if suitable tracts of land on the Shark Bay peninsula can be made secure and predator-free, it may be possible to re-introduce these endangered species and give them a second chance on the mainland..

Left: Banded Hare-wallaby *(Lagostrophus fasciatus)* . These animals seldom have access to fresh water and survive on moisture from heavy dew and food.

Below: Western Barred Bandicoot *(Perameles bougainville)* . At late dusk it emerges to forage for insects, seeds, roots and herbs obtained by hunting or digging.

Nature's Nurseries

Dirk Hartog Island provides the habitat of the rare White-winged Fairy Wren (found elsewhere only on Barrow Island). Other species of this spectacular group (the Splendid Wren and Purple Backed Wren) are common in Shark Bay.

The nature reserve islands of Shark Bay provide ideal nesting areas for at least fourteen species of sea-birds.

The Pied Cormorant rookeries on White Island and pelican rookeries on Pelican Island are a spectacular sight when the birds are nesting.

During the breeding season, visitors are asked to observe these birds from a distance. The birds are easily disturbed and a thoughtless person walking among the nests at critical times can cause damage to their breeding success.

Above: Pelicans *(Pelecanus conspicillatus)* are relatively tame and abundant in Shark Bay.

Top Right: In colonial days, Pied Cormorants *(Phalacrocorax varius)* formed the basis of the guano industry in Shark Bay. The bird's guano - excreta - high in phosphates and nitrogen made extremely good fertilizer.

Right: White-breasted Sea Eagle *(Haliaeetus leucogaster)*

Opposite Page: Domestic chores; White-winged Fairy Wren feeding her young.

WaterWonderland

Shark Bay supports a myriad of marine life, from whales, dolphins, dugongs, schools of manta rays and sharks to the smallest of hermit crabs and molluscs.

Marine turtles nest on remote northern beaches on Peron Peninsula and Dirk Hartog Island. Green Turtles *(Chelonia mydas)* and Loggerheads *(Caretta caretta)* predominate and are especially abundant on Green Turtle Flat, north of Faure Island and around the northern end of Peron Peninsula.

Hidden below the shallow waters of Shark Bay is one of the most species-rich and extensive assemblages of seagrasses in the world providing a habitat for a host of small animals.

Above: Loggerhead Turtle *(Caretta caretta)* on its slow journey back to the ocean after laying up to 100 eggs.

Right: Shrimp found in the coral reefs at Shark Bay.

The Wooramel Seagrass Bank along the eastern shore of Shark Bay, and the Faure Sill which bars the mouth of Hamelin Pool, are among the largest banks of carbonate sands in the world.

The shores throughout the bay are characterized by wide sand flats which are left bare at low tide. These fine sands provide habitats for a range of burrowing molluscs, crustaceans and other invertebrates.

Most of the species are widespread in tropical northern Australia, but there are many which are endemic to the west coast of Western Australia including the burrowing bivalve, *(Fragum erugatum)*, which forms the shell deposits of Hamelin Pool and Shell Beach.

Left: Shark Bay's extensive sand banks support a wealth of burrowing fauna and the most extensive assemblage of seagrass in the world.

Below Left: At Shell Beach, plunge into countless millions of tiny shells.

Below: Busy scavengers on the water's edge. Golden Ghost Crabs *(Ocypode convexa)*.

Delightful Dolphins

Imagine camping on a beach where dolphins come to the shore on their own free will to interact with humans. Ever seen the look in a wild dolphin's eye or touched its velvety skin?

At Monkey Mia dolphins come regularly to shore and allow themselves to be stroked (usually in the hope of fish in return!)

Every year, thousands of tourists flock to Shark Bay to see these Bottlenose Dolphins *(Tursiops truncatus)* in their natural environment.

The association between dolphins and people at Monkey Mia began many years ago. Locals say that it all began when "Old Charley" was handfed

Right and Below: A love affair begins when a human meets a dolphin.

from the side of a local fishing boat. He took the fish gracefully and continued to come back over following days.

The tradition appears to have been passed along to other members of the dolphin pod which live in the area.

Seven or eight dolphins are now regular visitors. Around 20 dolphins have been known to come to the jetty at various times and one of the calves is third generation.

They belong to a much larger group and one of the mysteries is why these particular individuals seek human contact.

A current, on-going research programme seeks information and understanding about the normal social relationships among the group, where the tame animals go when they aren't interacting with people, and what they do all day.

Bottlenose Dolphins are sometimes found roaming as solitary individuals or in groups ranging in size from two to hundreds. They have a fairly fluid social structure and groups seem to change composition with remarkable regularity.

Ther is no doubt that at Monkey Mia the dolphins love people and will happily play as long as a few simple rules are followed.

Dolphins object to the 'melon' on their forehead, their echolocation centre, being touched.

For visitors seeking information on the dolphins and the area there is a Dolphin Information Centre at Monkey Mia.

As part of a half million dollar programme to provide tourist facilities, the Centre was completed by the W.A. Government in conjunction with the Shire of Shark Bay.

Left: Living in harmony, openness and grace.

Below: Dolphins need respect.

Above and Right: Moments to remember. Two species communicate without words.

Because of the increasing pressure of crowds and the need to ensure the safety and welfare of the animals, rangers were introduced in 1986.

They are there to ensure the dolphins are not upset and that the animals don't become so overfed on handouts that they neglect their natural hunting activities.

Visitors are asked to observe the signs in the area and to consult the rangers in the proper methods of meeting these delightful mammals.

Gentle Giants

Probably the strangest of all marine mammals in the bay are the dugongs. Quiet, bottom feeding 'sea cows' that have been almost hunted to extinction in other parts of the Indian and Pacific Ocean.

The Shark Bay dugongs are believed to number about 1,000, representing one of the largest and most secure populations in the world. This is the southern extent of their range in Western Australia which spans from the Kimberley to Shark Bay.

During all seasons, dugongs occupy the warmest water available to them. Fluctuations imposed by water temperature variations see seasonal migration of herds of dugongs.

During summer, dugongs feed in the shallows of the Gladstone-Faure Sill area and in Freycinet Inlet. In winter, they move to warmer areas on the eastern side of Dirk Hartog and Bernier and Dorre Islands.

They graze on the dense growths of seagrass travelling between pastures in large groups though they are often found feeding alone. Their extraordinary snout is adapted for shovelling in the sand and extracting the choicest of plants.

These slow moving and gentle creatures are easily disturbed by power boats and visitors are asked to avoid the areas where they are feeding.

Below: Legends of mermaids arose from the first sighting of dugongs *(Dugong dugon)*. A dugong calf will stay with its mother for up to two years after birth.

Land of Living 'Fossils'

Stepping across the low sand dunes which surround much of Shark Bay, and descending to the beach, you could be forgiven for thinking that you had been transported billions of years back in time. Stretched out before you are domed, sedimentary structures called stromatolites.

Reaching about 30cm in height, these stromatolites are formed by tiny, single celled organisms called cyanobacteria. They construct the stromatolites by trapping fine sediment particles from the water and binding them together with a sticky film of mucus.

Stromatolites grow extremely slowly, generally less than 1mm per year, which means that those in Shark Bay are many hundreds of years old.

They are extremely vunerable to tampering. Footprints can be seen where tourists have walked and where wagon wheels carrying sandalwood and wool to ships at Shark Bay earlier this century crossed the intertidal flats. In these damaged areas, the stromatolites show little sign of regrowth.

In the areas of Shark Bay where the stromatolites grow the sea water is up to twice normal salinity. This is perhaps one reason for the stromatolites' survival here. The extra salty water may deter many of the fish and snails that otherwise might graze on them.

Below:This underwater view of stromatolites at Shark Bay resembles a scene 3.5 billion years ago when similar primitive life forms existed on earth.